CHEMICAL ELEMENTS

THE PERIODIC TABLE

The almost infinite objects and materials around us are actually made up of only a limited number of chemical elements. We know today that 91 exist naturally on Earth. They begin with hydrogen which was formed shortly after the universe came into existence. The other 90 were made either by nuclear reactions taking place in the core of burning stars or by the catastrophic explosions called supernovas that are sometimes produced when the stars die. Several more elements are made artificially in the laboratories.

Each element behaves differently and has different properties from all of the others. A system of organizing information about the chemical properties of the elements and the chemical compounds they form is essential. The modern periodic table is based primarily on the work of the Russian chemist Dmitry Mendeleyev whose table published in 1869 placed the elements in the horizontal rows according to their weight with one row beneath the other so that all the elements with similar properties fell into vertical columns. In the 20th century with knowledge gained about the structure of the atom ,the correct way of ordering the elements was discovered and the present periodic table was formulated.

Atoms made up of protons,neutrons and electrons are basic components of the elements. English physicist Henry Moseley demonstrated that what determines the behavior of each element is its atomic number ,the number of protons in its nucleus ,not its atomic weight which is a measure of the total number of protons and neutrons in the nucleus. The correct way of ordering the elements in the periodic table was therefore by their atomic number. Although the atoms of a given element have the same number of protons they can have different number of neutrons. These are called isotopes and their existence explains why the atomic weight is an unreliable indicator of the position of an element in the periodic table .

The elements are arranged in order of their atomic numbers in rows called periods. Moving from left to right across a period, there is transition of elements that are metals to those that are non-metals. The vertical columns of the periodic table are called groups. All the elements within a group have similar chemical properties and are sometimes referred to as families of elements.

WHY DO ELEMENTS WITHIN A GROUP HAVE SIMILAR CHEMICAL BEHAVIOUR

The atomic number determines how many negatively charged electrons are contained in the atoms of a particular element and it is the structure of the electrons orbiting the nucleus that determine how elements react with one another. This distribution of electrons in the valence ,or outer, shell of the atom are exposed to other atoms when they react. Elements whose valence shells are completely full are extremely stable and seem to react with almost nothing else. Those with incomplete shells will tend to react with other atoms in a manner that will complete these shells. Atoms with similar valence-shell configuration have similar chemical properties. Elements in the same group in the periodic table have the same number of valence electrons.

The periodic table then is a map of the way in which electrons arrange themselves in the atoms of a particular element. The ability to predict the chemical behavior of an element based on the row and column in which it is found makes the periodic table an invaluable reference tool for the practitioners of science.

HYDROGEN
Atomic number: 1
Chemical Symbol: H
Group:1A

Hydrogen consists of nothing more than a single proton, which serves as its nucleus, circled by a single electron. Its simplicity helps to explain why it is by far the most abundant element, making up 93% of all atoms in the universe. Hydrogen is a gas which has no odour or taste, is completely colourless -and extremely flammable.The combination of hydrogen with oxygen produces its most common compound,water.Hydrogen is also contained in organic compounds, biological compounds present in living organisms, in perfumes, dyes, pesticides, DNAs and proteins! The list goes on and on!

HELIUM
Atomic number:2
Chemical Symbol. He
Group VIII A-The noble gases

Like all noble gases, helium is colourless and odourless.Together hydrogen and helium form an astonishing 99.9% of elements in the universe. Its name comes from the Greek 'helios' which means the 'sun'. Helium from the sun is produced by the fusion of hydrogen. This reaction supplies the energy that the sun radiates into space. Helium has a low density and is therefore useful in blimps and toy balloons for its buoyancy in air.Astrnomers use the extremely cold liquid from of helium to remove thermal 'noise' making it easier and more reliable to receive data from distant galaxies.

LITHIUM
Atomic number: 3
Chemical Symbol: Li
Group IA-The Alkali metals

The metal lithium is extremely reactive and combines with aluminum to form low density ,structurally strong alloy used in aircrafts and spaceships. It is also used as a positive terminal or anode in small batteries used in cameras, pacemakers and calculators. Lithium hydroxide is a very efficient air-purifier. It absorbs CO_2 from the air to form lithium carbonate. Lithium has the highest heat capacity of any element. This property makes it ideal heat transfer material and it is being used in experimental nuclear reactors to absorb the heat produced by the fissioning of uranium.
In medicine lithium carbonate and lithium citrate are known as very effective mood stabilizers in manic-depressive illness.

BERYLLIUM
Atomic number:4
Chemical Symbol: Be
Group IIA-The Alkaline Earth Metals

In its pure form, Beryllium is a light, fairly hard, gray-white metal. Like all metals that make up the alkaline earth group, it is much too chemically reactive to be found in its free state. Deposits of the mineral beryllium are distributed over Brazil, Argentina, and the US. Crystals of beryllium are known for their exquisite appearance. Both emerald and aquamarine are naturally occurring precious forms of this mineral. Beryllium played a key role in the discovery of the neutron in 1932 and remains useful in researches on atomic nuclei.

BORON
Atomic number:5
Chemical Symbol: B
Group III A

Boron is a hard ,brittle, non-metallic element. It is usually bound with oxygen ,water and sodium in a compound called borax that is used as a cleaning agent and water softener. When water is softened, the magnesium and calcium are replaced with relatively harmless sodium and Potassium. Another boron compound is boric aced used industrially to make Pyrex, a special heat resistant glass used in kitchens. Boron 'rods' are crucial in the utilization of nuclear reactors. They can be lowered into a reactor to absorb neutrons thus controlling the power being produced by the reactor.

CARBON
Atomic number: 6
Chemical symbol: C
Group IV A

Carbon represents only 0.09% of the earth's crust by mass ,yet it is the element most essential for life on our planet. Carbon owes its central position in the organic world to the ability of its atoms to link up with other carbon atoms to form long chains that are either straight or branched. One such long chained molecule in the DNA found in the genetic material of all living creatures. Elements can exist in several natural forms called allotropes. Carbon is found in the allotropic forms of graphite, coal and most spectacularly diamond.

NITROGEN
Atomic number: 7
Chemical symbol: N
Group V A

Nitrogen lacks any sense stimulation property and we are constantly breathing in large quantities as we inhale air. It dominates the gases in earth's atmosphere making up some 78% by volume. Nitrogen forms hundreds of thousands of compounds that are crucial for agriculture and industry the most important of which is ammonia. In its gaseous form, nitrogen is often used in situations in which it is important to keep other, more reactive atmospheric gases away. For example, to prevent the oxidation of wine ,wine bottles are often filled with nitrogen after the cork is removed.

OXYGEN
Atomic number: 8
Chemical symbol: O
Group VI A

Oxygen exists in the atmosphere in water, and in the earth's crust in an enormous variety of rocks and minerals. It is essential for life and part of every biological molecule in our bodies. Although many natural processes consume oxygen, it is constantly replenished by photosynthesis in plants thus continually being consumed and continually being produced. The English chemist Joseph Priestley is credited with the discovery of oxygen. He heated an oxide of mercury and noted that the gas it gave off caused the candle to burn with a remarkably brilliant flame. The gas was oxygen!

FLUORINE
Atomic number:9
Chemical symbol: F

Group VII A-The Halogens
Fluorine is the smallest, lightest and the most reactive halogen. All atoms in this group readily combine with metals to form salts. In many parts of the world sodium fluoride is added to public water supplies. Research has shown that small quantities of fluorine can retard the development of cavities in teeth. In the presence of hydrogen, fluorine burns with explosive force producing hydrogen fluoride which when dissolved in water forms hydrofluoric acid. It is extremely dangerous. However ,it is used to dissolve glass and is used to etch design on glass objects.

NEON
Atomic number:10
Chemical symbol: Ne
Group VIII A-The Noble Gases

Neon like all noble gases is monoatomic. The familiar neon signs in storefront and restaurant windows contain neon gas that glows when it is energized by an electrical discharge. When this happens, neon atoms in the gas give off radiation in the form of orange-red light. Different gases are used to produce signs of different colurs. Every gas when excited radiates its own characteristic colour. Commercial neon is produced in air-liquefaction plants. Because neon has a boiling point of -229 degree Centigrade ,it remains as a residue after the more volatile nitrogen and oxygen have boiled off!

SODIUM
Atomic number: 11
Chemical symbol: Na

Group IA-The Alkali Metals

Sodium is an extremely reactive bright silvery metal light enough to float on water and soft enough to be cut with knife. It is a part of many important compounds that are found widely distributed throughout the earth. Sodium chloride, the chemical name for table salt is mined in huge quantities from natural salt deposits. Sodium bicarbonate commonly known as baking soda is used to make baked goods rise when heated or pastry dough rise when baked. It is also used to neutralise excessive stomach acidity and as an agent in fire extinguishers.

MAGNESIUM
Atomic Number: 12
Chemical symbol: Mg
Group II A-The Alkaline Earth Metals

Magnesium is present in such large quantities in seawater that the world's oceans contain an almost unlimited supply of the dissolved material. Its great advantage is that it is very light which also makes it ideal for fabricating automobile and aircraft parts, power tools, lawn mower housings and racing bikes. Magnesium is also important for proper nutrition in humans because it is essential for proper functioning of several enzymes. It also plays a crucial role in the make-up of the green chlorophylls present in all green plant cells.

ALUMINUM
Atomic number: 13
Chemical symbol: Al
Group III A

Usually found in nature combined with oxygen, aluminum is the most abundant metal in the earth's crust. It is lightweight and good conductor of electricity, two properties that make it an ideal ingredient for a wide range of products. It is an excellent reflector of radiation and is used for various types of antennas, heat reflectors, and solar mirrors. Beyond these other properties, aluminum is fairly reactive. It forms an oxide layer that prevents it from further reactions with the environment so that it is usually considered corrosion-resistant. Aluminum is also non-toxic, odourless and tasteless.

SILICON
Atomic number: 14
Chemical Symbol: Si
Group IV A

Compounds of silicon bound chemically to oxygen make up most of the earth's sand, rock and soil. Today silicon forms the basis of microelectronics industry. The use of silicon chips in printed circuits has made it possible the shrinking room sized computers into ones that can rest on your lap. The most important silicon compound is silica which exists in two forms- quartz and flint. Small gems and semi-precious stones are crystals of quartz with coloured impurities. Silica is used in the production of glass. Ceramics and silicones are other important classes of compounds based on silicon.

PHOSPHORUS
Atomic number:15
Chemical symbol: P
Group VA

Phosphorus was discovered by physician Hennig Brand in 1669. He distilled the residue from boiled down urine and obtained something that glowed in the dark and burst into flames in warm air. Phosphorus and light emission are still linked in the phenomenon known as phosphorescence. Zinc sulfide is the phosphorescent material that gives off scintillations of light when struck by fast moving electrons. This effect on the coating of television tube produces the TV image. Almost all phosphorus used commercially is to make phosphoric acid. Its major use is in the production of fertilizers-soil without phosphorus is barren. Commonly found in two forms i.e. red and yellow, the former is used to make safety matches.

SULPHUR
Atomic number: 16
Chemical symbol: S
Group VI A

Sulphur is a reactive non-metal found in nature both in its free elemental state and in the form of widely distributed ores and minerals. Some common minerals of Sulphur are gypsum i.e. calcium sulfate and pyrite often known as the 'fool's gold'. In addition to their importance in making artificial fertilizers, preserving food, bleaching textiles and cleaning metals ,Sulphur compounds have hundreds of other uses in recovering metals from ores, making rubber, detergents, paints and dyes, and synthetic fibers. Indeed a nation's level of industrial development is determined by its per capita consumption of Sulphur.

CHLORINE
Atomic number: 17
Chemical symbol: Cl
Group VII A-The Halogens

Chlorine is a poisonous yellowish green diatomic gas. Inhaling even a small amount can cause serious lung damage. The toxicity of chorine makes it an excellent disinfectant for swimming pools and water supplies. An important compound of chlorine is hydrogen chloride ,a gas that dissolves in water to produce hydrochloric acid. Hydrochloric acid is present in the gastric juice of stomach where it is needed to activate protein digesting enzymes. Large amounts of chlorine have been used to produce insecticides. Many have been recently banned as they are regarded as environment pollutants.

ARGON
Atomic number:18
Chemical symbol:Ar
Group VIII A-The Noble Gases

In 1894, argon became the first noble gas to be discovered. Its commercial applications make use of its lack of reactivity. Argon is the decay product of an important radio-isotope used for dating rock samples, potassium-40.The technique is called potassium-argon dating. Potassium has an unusually long half life of 1.25 billion years and is present in many rocks. When potassium 40 decays, it transforms itself into argon. Consequently one can determine the age of a rock by determining how much argon is present. The oldest rocks on earth have been determined by this method as 3.8 billion years old.

POTASSIUM
Atomic number: 19
Chemical Symbol: K
Group IA The Alkali Metals

Potassium is extremely reactive hence is never found in its free state in nature. It is found in sea-water ,although in smaller amounts than sodium, its chemical equivalent. Potassium is essential for plant growth so much of the potassium in dissolved minerals is taken up by plants before reaching the sea. A naturally occurring isotope of potassium is potssium-40.Human body contains 140 grams of potassium. Since the abundance of potassium-40 is 0.012 percent, we are all partially made up of this reactive isotope. It is a major contributor to our lifetime dose of radiation

CALCIUM

Atomic number: 20
Chemical Symbol: Ca
Group II A-The Alkali Earth Metals

Calcium is an important ingredient for a wide range of living organisms. Human teeth and bones contain calcium and marine organs build their shells of calcium carbonate. Lime, a compound of calcium is an essential industrial chemical. One of its early uses was in theatrical lighting. When lime is heated to a high temperature, it gives off an intense bluish-white light. It was used in early 19th century to illuminate actors giving rise to the phrase 'in the limelight.' Probably the most important modern use of lime is in the production of iron from its ores.

SCANDIUM
Atomic number: 21
Chemical Symbol: Sc
Group III B First Row Transition Element

Scandium heads the first row transition elements. All are fairly unreactive metals and many are extremely hazardous. Scandium is a very light weight metal with a fairly high melting point and shows good resistance to corrosion. These properties have made it of great interest to the aerospace industry for construction of an aircraft. Scandium forms few useful compounds. The metal itself has found some use in electronic devices such as high intensity lamps that produce light with a colour value close to that of natural sunlight. Lamps of this kind are often used to illuminate soccer stadiums.

TITANIUM
Atomic number: 22
Chemical symbol: Ti
Group IV B First Row transition Element

Titanium in its pure state is a metal that is easy to work and quite ductile or capable of being drawn into wire. Despite its light weight ,it is unusually strong and virtually immune to usual kinds of metal fatigue. It also has an extraordinary resistance to corrosion so that it has every property needed to make it an ideal material for jet engines and rockets. The most important compound is titanium dioxide a substance with intense brilliant white colour that is used as a pigment for paints, paper and plastic.

VANADIUM
Atomic number: 23
Chemical symbol:V
Group VB First Row Transition Element

Vanadium is a bright shiny metal that is fairly soft and extremely resistant to corrosion. A Mexican professor of mineralogy viz Andres Manuel del Rio discovered vanadium in 1801. It was later named after the Scandinavian goddess Vanadis because of its many beautifully colored compounds. About 80 % of the vanadium produced in the US goes into the manufacture of steel.

CHROMIUM
Atonic number: 24
Chemical Symbol: Cr
Group VI B First Row Transition Element

Chromium was named from the Greek word 'chroma' meaning colour. The beautiful colour of many precious gems -the red of rubies, the characteristic green of the emeralds -is owing to the presence of trace amount of chromium. The metal is usually extracted from chromite, an oxide of chromium that is its most important ore. When exposed to air, chromium forms an invisible oxide that makes it extremely resistant to corrosion and very useful both as a decorative and protective coating over other metals such as brass, bronze and steel. Chromium is also used to produce stainless steel.

MANGANESE
Atomic number: 25
Chemical symbol: Mn
Group VII B First Row Transition Element

Manganese is a hard gray-white metal that looks like and has many properties similar to iron. Adding manganese to steel makes is unusually hard and resistant to shock. Such steel is ideal for use in rifle barrels, bank vaults, railroad tracks, and earth moving equipment. Manganese also adds hardness, strength and corrosion resistance to alloys of aluminum and magnesium. The compound potassium permanganate has a purplish colour that is sometimes seen in antique glass. Although glass manufacturers no longer use manganese, its ability to colour objects is used to brighten ceramics and pottery.

IRON
Atomic number: 26
Chemical symbol: Fe
Group VIII B First Row Transition Element

Iron is probably the most common metal in the human society. Whether we are using a screwdriver or riding a car or a train, the importance and usefulness of iron as a structural material is self evident. The interior of the earth known as core is made of molten iron. The ability to refine the metal served as a major milestone in human development known as the Iron Age (1000 BC). Its discovery lead to tools and weapons that were harder and more durable than those of Bronze Age. Today more than 90% of all metals refined is iron.

COBALT
Atomic number: 27
Chemical symbol: Co
Group VIII B First Row Transition Element

A major ore of cobalt is cobaltite. The pure metal is obtained by roasting this ore. The name cobalt comes from the German 'kobold' which refers to an evil spirit. Miners often said that accidents occurring in the mind were caused by 'kobold'. Cobalt is added to steel to improve its resistance to corrosion. When cobalt is mixed with tungsten and copper, it forms Stellite, a metal that retains its hardness at high temperatures making it ideal for high speed drills and cutting instruments. Like iron cobalt is easily magnetized. The powerful magnetic substance known as alnico is an alloy of cobalt, aluminum and nickel.

NICKEL
Atomic number: 28
Chemical symbol: Ni
Group VIII B First Row Transition Element

Nickel is frequently added to other metals such as iron and steel to form alloys resistant to oxidation. Nichrome the metal used to make the heating elements in toasters and electric ovens is an alloy of chromium and nickel. The high electric resistance of nichrome combined with its high melting point makes it a very efficient material to convert electricity to heat. An important use of the metal is in nickel-cadmium batteries. This battery is rechargeable which makes it particularly useful in calculators, computers and cordless electric shavers.

COPPER

Atomic number: 29
Chemical symbol: Cu
Group IB First Row Transition Element

A familiar use of water is in the pipes that carry the water into the kitchen. Because copper is one of the best conductors of electricity ,copper wires are widely used to transmit electrical energy from power stations to homes, offices ,factories and other buildings and from wall outlets to electrical appliances. Copper was once used to make buttons for uniform jackets for policemen hence the colloquial 'copper' for the police. Brass, an alloy of copper and zinc has a wide variety of uses from hardware to zinc.

ZINC
Atomic number: 30
Chemical symbol: Zn
Group I B First Row Transition Element

In its pure state ,zinc is a hard, brittle ,silvery metal. It is relatively corrosion resistant and quickly forms a hard oxide coating that prevents it from reacting further with the air. In the process called galvanization ,a layer of zinc is coated over steel to prevent corrosion. The metal has many other uses. One of the most important is in the common dry cell battery. Since 1981 zinc has served as the chief metal in the US penny. Zinc is also combined with copper to form brass.

GALLIUM
Atomic number: 31
Chemical symbol: Ga
Group III A Post Transition Metal

Gallium is an extremely soft metal with a very low melting point and an extremely high boiling point of 2403 degree Centigrade. The range of temperatures at which gallium is liquid is the largest of any known metal. This makes it useful for special high degree thermometers .Until recently few practical applications of gallium were known. This changed rapidly with the discovery that gallium arsenide could function as a laser diode and convert electricity directly into laser light. Light emitting diodes are used in a variety of watches and autodisc players.

GERMANIUM
Atomic number:32
Chemical symbol:Ge
Group IV A Metalloid

Germanium is a relatively rare dark gray solid element. It is never found in pure form in nature but combined with oxygen. Germanium is called a semi-conductor. The addition of small amount of impurities greatly increases its capacity to conduct electricity.'Doped' germanium is used to make transistors that are at the heart of the solid state electronics industry. With doping tens of thousands of transistors can now be formed on a small germanium chip which in effect becomes a small computer. Such materials have made possible the revolution in electronics miniaturization.

ARSENIC
Atomic number: 33
Chemical symbol: As
Group VA Metalloid

Arsenic is a brittle crystalline solid at room temperature. In the form of arsenious oxide it is a well known poison. It is used as a weed killer and insecticide. Arsenic as poison has captured the imagination of many a crime writer. Before recent advances in forensic techniques, it was impossible to detect in the victim's body. Although a poison, arsenic compounds have been used for medicinal purposes as well, the most well known being '606' devised by Paul Ehrlich as a cure for syphilis.

SELENIUM
Atomic number: 34
Chemical symbol: Se
Group VI A Metalloid

Selenium bearing minerals are too scarce to be mined profitably. Because the metalloid is found in the company of copper and Sulphur ,almost all selenium is recovered as a bye-product of copper refining and the manufacture of sulfuric acid. Selenium exists in two forms-red and gray. Gray selenium is a photoconductor meaning that although a poor conductor of electricity ordinarily, it becomes and excellent conductor in presence of light. This makes selenium valuable as a light sensor in robotics and light meters.

BROMINE
Atomic number: 35
Chemical symbol: Br
Group VII A The Halogens

Bromine is a reddish liquid with an acrid smell. Its name is derived from the Greek bromos meaning stench. Bromine can be found in seawater, underground salt mines, and deep brine wells. A major use of bromine is in producing a petrol additive called ethylene dibromide. This compound removes the lead additives after the combustion of petrol preventing the formation of lead deposits. Bromine is extremely toxic and burns the skin. Moreover its noxious vapours can damage nose and throat.

KRYPTON
Atomic number:36
Chemical symbol:Kr
Group VIII A The Noble Gases

In 1933 Linus Pauling challenged the idea that the noble gases were chemically inert. The existence of the compound he predicted of krypton and fluorine was confirmed in 1966. Krypton is an odourless, tasteless, colorless completely harmless gas .Its chief use is in 'neon' lights that are a part of the modern landscape. When sealed in a glass tube and subjected to electrical discharge ,krypton produces a pale violet colour used for airport runway and approach lights. Krypton is also used mixed with xenon in high intensity, short-exposure photographic flash bulbs or strobe lights.

RUBIDIUM
Atomic number: 37
Chemical symbol: Rb
Group IA The Alkali Metals

Rubidium is a silvery,very soft highly reactive metal that burns spontaneously when exposed to air. It also reacts violently with water giving out large quantities of hydrogen that immediately bursts into flames because of the heat generated by the reaction. Rubidium is much too reactive to exist as pure metal in nature and few rubidium bearing minerals are known. Rubidium has little commercial value. The metal was discovered in 1861 by German chemists Robert Bunsen and Gustav Kirchoff. They identified it by spectral lines as an impurity among many alkali metals they were investigating.

STRONTIUM
Atomic number: 38
Chemical symbol: Sr

Group IIA The Alkaline Earth Metals

Strontium has little commercial use and its compounds have found only limited application in industry. Since strontium salts such as strontium carbonate emit a characteristic red colour when they burn, they are used in highway warning flares and in fireworks. One of the isotopes of strontium, Sr-90 is a radioactive by product of nuclear explosions and can contaminate large areas of environment through fallout from the atmosphere. Since strontium 90 is produced whenever uranium undergoes fission, operators of nuclear reactors must be constantly on guard to prevent its accidental release into the environment.

YTTRIUM
Atomic number: 39
Chemical symbol: Y
Group III B Transition Element

Yttrium is found in small quantities in the earth's crust but the rocks brought back from the Moon had an unexpectedly high yttrium content. When their temperature is lowered to only a few degrees above absolute zero, almost all metals show no electrical resistance whatsoever. Extremely low temperatures are impractical however. In 1987 scientists announced the discovery of a compound of yttrium, copper and barium oxide that was superconducting at 93 degrees Kelvin . Other mixtures of this element are being investigated and there is optimism that one of them would turn out to be a practical high temperature superconductor.

ZIRCONIUM
Atomic number: 40
Chemical symbol: Zr
Group IV B Transition Element

Zirconium is a strong, durable metal. Its ability to withstand high temperatures makes it an ideal ingredient for heat resistant materials in the spacecraft. The best known compound of zirconium is the metal zircon .It has been known since ancient times and even referred to in the Bible. Found in a wide variety of colours, when the crystal is cut and polished it is regarded as a semi precious gem. Zircon has an extremely high index of refraction. Because of this, its colourless crystals have an unusual brilliance and are sometimes used as substitutes for diamonds.

NIOBIUM
Atomic number: 41
Chemical symbol: Nb
Group VB Transition Element

The metal niobium has been important in the history of high temperature superconductivity. An alloy consisting of niobium and germanium has the ability to withstand large currents permitting the construction of superconducting magnets for such instruments as nuclear magnetic
resonance scanners used in diagnostic medicine. Niobium is added to steel for special purposes. At high temperatures the boundaries between the small grains that make up stainless steel weaken and corrode more easily than the rest of the steel. The addition of niobium prevents this from happening allowing steel to withstand much higher temperatures under extreme stress.

MOLYBDENUM
Atomic number: 42
Chemical symbol: Mb
Group VI B Transition Element

Molybdenum is a hard silvery metal. Fairly large deposits of molybdenite are found in Colorado ,US. Steel containing molybdenum is well suited for aircraft and car engine parts. It is able to withstand temperature and pressure changes constantly taking place in an engine. For the same reason it is used in the manufacture of guns and cannons. One of the radioactive isotopes,molybdenum-99 is used in hospitals to generate technetium-99 which is highly useful for taking pictures of internal organs after being taken internally.

TECHNETIUM
Atomic number:43
Chemical symbol: Tc
Group VII B Transition Element

Technetium was the first element to be produced in laboratory from another element.Logically it takes its name from the Greek teknetos meaning artificial. Every isotope is radioactive and decays to form an isotope of a different element. Today nuclear reactors produce one of the most useful isotopes of technetium ,technetium-99m. When it in injected into the veins of a patient, the isotope will concentrate in certain body organs and its radioactivity will expose a photographic plate revealing how those organs are functioning.

RUTHENIUM
Atomic number: 44
*Chemical symbol:*Ru
Group VIII B Transition Element

Ruthenium is a rare element that is usually recovered as a by product of the refining of platinum ores. Mainly ruthenium is used as a catalyst for industrial processes. It has been used as a catalyst in obtaining hydrogen gas directly splitting water molecules rather than by electrolysis.Rutheniumis also used in the jewellery business as a hardening additive to platinum and is often added to titanium to improve its resistance to corrosion. Other alloys of ruthenium are used in fountain pen points and special electrical contacts.

RHODIUM
Atomic number: 45
Chemical symbol: Rh
Group VIII B Transition Element

Rhodium is a rare, extremely hard silvery gray metal .It was discovered by William Wollaston in 1803. He named it after the Greek word rhodon for rose because many of the salts have rose colour. It is used in the catalytic convertors of cars. The exhaust gases are a major source of atmospheric pollution. The catalytic convertor is filled with small catalytic beads containing platinum, palladium and rhodium which convert hot exhaust gases that pass through them into harmless products.

PALLADIUM
Atomic number: 46
Chemical symbol: Pd
Group VIII B Transition Element

Palladium is a soft silvery white metal that resembles platinum. It is extremely malleable and ductile. An interesting use of palladium emerged when it was serendipitously determined that it was useful in treating cancers by inhibiting cell division and was relatively free of side effects. With a half life of only 17 days ,the palladium103 isotope can deliver powerful doses of radiation to destroy cancer and then disappear after a little more than a month.

SILVER
Atomic number: 47
Chemical symbol: Ag
Group IB Transition Element (Coinage Metal)

Silver is one of the few metals found in free state in nature and its symbol Ag comes from Latin word argentum which means silver. It has been a coinage metal since Biblical times maybe even earlier. Of all metals, silver is the best conductor of heat and electricity. It is not usually used in home wiring because of expense but extensively used in the manufacture of high quality electronic equipment.

CADMIUM
Atomic number:48
Chemical symbol: Cd
Group II B Transition Element

Cadmium is present in such great quantities of zinc ores that it is generally considered a by product of zinc refining. The major use of the metal is in electroplating of steel to prevent it from corrosion. It is used less often than zinc because it is less abundant and has a propensity to cause health problems. The ability of cadmium to absorb neutrons is of great importance in the design of nuclear reactor control rods. Cadmium is also used as a red and yellow pigment in making paint.

INDIUM
Atomic number: 49
Chemical symbol: In
Group III A Post transition metal

Indium is a rare bluish white metal soft enough to leave traces of itself when vigorously rubbed against other metals. Pure indium has few commercial applications and it is mainly used as an alloy with other metals. Alloys of indium and silver and indium and lead are better conductors than silver or lead alone. They have also found uses in manufacture of transistors and photo cells. Indium foils are often inserted into nuclear reactors to control the nuclear reaction. The rate at which these foils become radioactive serves as a valuable measurement of the reactions taking place.

TIN
Atomic number: 50
Chemical symbol: Sn
Group IV A Post Transition Metal

Tin was among the first metals used by human beings. Bronze, an alloy of copper and tin was used in Egypt more than 5000 years ago. Today it is mainly used as an alloying agent and to make tin plate which is steel sheeting covered with a thin coating of tin. Because tin protects steel from food acids , tin plate was used to make tin cans for food but has now been largely replaced by plastic and aluminum. It is one of the most malleable metals known.

ANTIMONY
Atomic number: 51
Chemical symbol: Sb
Group VA Metalloid

Antimony is a hard,brittle,crystalline,grayish,solid. Although known as a metal, it is a very poor conductor of electricity. The ore that serves as the primary source is the mineral stibnite. A black compound, it was used in ancient times to darken women's eyebrows. A major use for the antimony is common safety match. The head of the matchstick contains a mixture of antimony trisulfide and an oxidising agent such as potassium chlorate. Antimony has few other commercial uses. As an alloy it can increase the hardness of many metals.

TELLURIUM
Atomic number: 52
Chemical symbol: Te
Group VI A Metalloid

Tellurium is a rare silvery-white metalloid. Unlike typical metals, it is brittle and a poor conductor of electricity. Tellurium is one of the few elements that combines with gold. The compounds it forms are called gold tellurides and they make up a very important component of gold bearing ores. Tellurium is often recovered as a by product in refinement of gold and also of copper. The chief use of tellurium is as an additive to such metals as copper and stainless steel to create an alloy that is easier to machine than the original metal.

IODINE
Atomic number:53
Chemical symbol:I
Group VIIA The Halogens

Iodine is a violet black solid found in seaweeds ,brine wells and in the sea. Although a poison, one of its commonest uses is as an antiseptic solution tincture of iodine. Iodine salts are added to table salt and animal feed. This is done as iodine is an important constituent of the hormone thyroxine secreted by thyroid glands and helps ensure that the gland functions properly. Silver iodide has the ability to form enormous number of crystals -as many as one million billion from one gram-which act as nuclei for raindrop formation.

XENON
Atomic number; 54
Chemical symbol: Xe
Group VIII A The Noble Gases

Xenon exists in atmosphere in only trace amounts. Like the other noble gases it exists as a monoatomic molecule that has no colour odour or taste. In 1962,Neil Bartlett the English chemist made the first noble gas compound. He combined xenon and platinum hexafluoride and much to his astonishment obtained a solid, yellow-orange compound which consisted of molecules of xenon,platinim and fluorine. To date xenon and krypton are the only noble gases known to form compounds. Like other noble gases, xenon is used in electrical discharge tubes to produce light.

CAESIUM
Atomic number: 55
Chemical symbol: Cs
Group IA The Alkali Metals

Pure cesium is the softest metal known. Its extreme reactivity has made it useful in removing unwanted gases from vacuum systems for example inside a television tube. The isotope caesium-133 serves as the world's official measure of time. The second is measured in terms of the radiation emitted by caesium 133 atom when it is excited by an external energy source rather than in terms of the earth's rotation around the sun as it used to be . The second is described as the elapsed time of exactly 9,192,531,770 vibrations of the radiation emitted by caesuim-133 atom.

BARIUM
Atomic number: 56
Chemical symbol: Ba
Group IIA The Alkaline Earth Metals

In the form of soluble salt, barium is quite toxic. On the other hand in insoluble forms it is harmless to the human body. Radiologists use barium sulfate to examine a patient's intestinal tract with Xrays. Barium sulfate has also a number of other uses based on its low solubility in water and white colour. It is used as a whitener on photographic plates and as a filler in writing paper, plastics and artificial fibres. Barium metal has few commercial applications because of its readiness to react with oxygen and moisture.

LANTHANUM
Atomic number: 57
Chemical symbol: La
Group III B Rare Earth Element (Lanthanides)

Lanthanum is the first of the rare earth element series. It is common to find many rare elements mixed together in a single mineral. Probably the most important use of lanthanide compounds is in fabricating the electrodes for the high intensity carbon arc lamps used in searchlights ,studio lighting and motion picture projectors. Lanthanum and its isotopes are found in the fragments that are produced when uranium fissions. It was the discovery of lanthanum isotopes as well as those of barium by German chemist Otto Hahn that eventually lead to the idea of nuclear fission.

CERIUM
Atomic number: 58
Chemical symbol: Ce
Group III B Rare Earth Elements (Lanthanides)

Cerium was named after the asteroid Ceres whose discovery in 1801 caused great excitement in the scientific world. The pure metallic form of cerium was not prepared until 1875. It is an iron gray metal that is quite malleable and ductile. Cerium compounds like those of lanthanum are used commercially to form electrodes of the high intensity carbon arc lamps .As an oxide cerium is used as an additive to the walls of self-cleaning ovens where it seems to prevent the buildup of cooking residues.

PRASEODYMIUM
Atomic number: 59
Chemical symbol: Pr
Group III B Rare Earth Elements (Lanthanides)

It was discovered by Carl Auer von Welsbach ,an Austrian baron who had an interest in mineralogy. The pure metal is isolated from its ores by ion-exchange technique. An exchange process is used to isolate one kind of ion by substituting it with another. In one such process the active ingredient is a resin made up of large molecules that have a netlike structure. The resin contains mobile ions loosely connected to the net. When a solution containing the other ions is passed through the resin ,they replace the mobile ions that then diffuse out of the net.

NEODYMIUM
Atomic number: 60
Chemical symbol: Nd
Group III A Rare Earth Elements (Lanthanides)

It is a magnetic substance used to create some of the most powerful magnets in the world. The supermagnets are known as NIB magnets as they contain iron and boron as well. They are so strong that two small magnets with press to either side of one's hand without falling. A Nd magnet with only half inch diameter is strong enough to respond to magnetic materials in printing ink used in paper money and can be used to detect counterfeit. It is also used in rose coloured glasses!

PROMETHIUM
Atomic number: 61
Chemical symbol: Pm
Group III B Rare Earth Elements (Lanthanides)

No trace of promethium has been found on the Earth's crust but it has been identified in the spectrum of several stars in the Andromeda Galaxy. It is a synthetic rare element made in the nuclear accelerators and nuclear reactors. When neodymium is subjected to the intense neutron radiation present in a reactor ,it is converted into promethium. 28 isotopes of the element have so far been synthesized all being radioactive. Very little is known of the chemical and physical properties of pure promethium.

SAMARIUM
Atomic number: 62
Chemical symbol, Sm
Group III B Rare Earth Element (Lanthanides)

The principal ores of samarium are bastnasite and monazite. Monazite ores often containing as much as 50% of their weights in rare earths are found in river sands in India and Brazil and in Florida beach sand.In its pure form samarium has a silvery-white luster and is fairly resistant to oxidation. The metal will however ignite spontaneously at low temperatures. Some compounds of this element are used to fabricate permanent magnets. Samarium oxide is an excellent absorber of infra-red radiation and is added for this purpose to various types of glass and infrared sensitive phosphorus.

EUROPIUM
Atomic number: 63
Chemical symbol, Eu
Group III B Rare Earth Element (Lanthanides)

Europium is one of the rarest of the rare earth metals. In 1901 French chemist Eugene-Anatole Demarcay finally isolated an impurity in a samarium-gadolinium sample he was studying and identified the impurity as a new element. Pure europium is fairly soft and silvery white. It is quite ductile and one of the most reactive of the rare earth metals. Europium oxide is fairly widely used as an additive to improve the efficiency of red phosphor in television and computer monitors. It is also used to increase the energy efficiency of fluorescent lamps.

GADOLINIUM
Atomic number: 64
Chemical symbol: Gd
Group IIIA Rare Earth Element(Lanthanides)

Two isotopes of gadolinium are among the most potent absorbers of neutrons. Though their scarcity limits use, they are used in making control rods for nuclear reactors. It is ferromagnetic meaning that it is very strongly attracted by magnets. However its Curie point, the temperature at which magnetic material loses its magnetism is approximately room temperature. It has been proven of value in a technique probing the interior of metals called neutron radiography. It is used in the airline and ship building industries to search for hidden flaws and structural weaknesses in hulls and fuselages.

TERBIUM
Atomic number: 65
Chemical symbol: Tb
Group III B Rare Earth Element (Lanthanides)

In a pure metallic form, terbium is a silvery-white, malleable, ductile and soft enough to be cut with a knife. It bears a resemblance to lead but it is much heavier. Like lead it is fairly resistant to corrosion. Compounds of terbium have founds uses in special lasers and as phosphors that produce the green colour in television tubes and computer monitors. Other applications include the production of alloys with special magnetic properties for use in compact discs and in the fabrication of high definition X-ray screens.

DYSPROSIUM
Atomic number: 66
Chemical symbol: Dy
Group III B Rare Earth Element (Lanthanides)

Dysprosium ranks ninth in abundance among the rare earth elements in the Earth's crust. It was discovered in 1886 by French chemist Paul-Emile Lecoq de Boisbaudran in a sample of erbium oxide. He based its name on the Greek word dysprositos which means hard to get at. Pure dysprosium was not available until 1950 when modern chemical techniques such as ion-exchange separation were developed. Dysprosium resembles most of the other rare earth metals. It is soft enough to be cut with a knife, has a shiny silvery colour and is relatively stable in air.

HOLMIUM
Atomic number: 67
Chemical symbol: Ho
Group III B Rare Earth Element (Lanthanides)

In 1878, two Swiss scientists noticed holmium's characteristic spectral lines but could not identify them. They called the unknown source of the spectral lines element X. Soon afterwards in 1879 Swedish chemist Per Teodor Cleve isolated and identified the element while working with a mineral called erbia. Pure metallic holmium which was not available until quite recently has a bright silvery colour. It is fairly corrosion resistant in dry air but tarnishes quickly in moist air forming a yellowish oxide. Other than its use as a colour for glass, it has few commercial applications.

ERBIUM
Atomic number: 68
Chemical symbol: Er
Group III B Rare Earth Element

Erbium was discovered by Carl Gustaf Mosander in a yellow oxide that he isolated from the mineral yttria. Mosander named the element for the Swedish village of Ytterby the site of large concentrations of yttria and erbium. The principal sources of erbium are the minerals xenotime and euxerite. Erbium as well as other rare earth elements is actually an impurity in these ores. The commercial applications of erbium are rather limited. Its oxides are often added to glass and enamel glazes to colour them pink. The glass is often used for sunglasses and inexpensive jewelry.

THULIUM
Atomic number: 69
Chemical symbol: Tm
Group IIIB Rare Earth Element (Lanthanides)

Thulium is a rare earth element that is extremely scarce. It occurs in very small quantities in the company of other rare earths. The Swedish chemist Per Teodor Cleve discovered the element in 1879 and named it for Thule, the ancient name for Scandinavia. The principal source of thulium is the mineral monazite which consists of approximately seven thousandths of 1% thulium. It has few commercial applications apart from being used in lasers. It is expensive but very little of the metal is available for experimentation.

YTTERBIUM
Atomic number: 70
Chemical symbol: Yb
Group III B Rare Earth Element (Lanthanides)

Ytterbium, the first rare element to be discovered is found in modest abundance in the Earth's crust and always in company of rare earths. It was discovered by the French chemist Jean de Marignac in 1878 as a component of the mineral known as erbia and named for the Swedish village Ytterby on the basis of its high concentrations of erbium. Pure ytterbium metal was not available for study until 1953. Its commercial applications are as an alloying agent with stainless steel. Certain alloys have also been used in dentistry.

LUTETIUM
Atomic number: 71
Chemical symbol: Lu
Group III B Rare Earth Element (Lanthanides)

Although he never formally published his results, US chemist Charles James is now considered to have discovered lutetium in 1907. Working during the early 1900's at the University of New Hampshire, James became a major force in the production of rare earth elements. He and his students would process tons of ore and labour through crystallizations to produce a single sample. Pure lutetium metal is difficult and expensive to prepare. It is the hardest and the heaviest rare earth element. No commercial applications have been developed.

HAFNIUM
Atomic number: 72
Chemical symbol: Hf
Group IV B Transition Element

Hafnium's properties as well as its history are closely tied to zirconium. Many had predicted the existence of element 72 but the omnipresence of its chemical twin interfered with its identification. The principal use of hafnium is based on one of its few differences from zirconium. Its ability to absorb thermal neutrons makes it a useful material for reactor control rods. The main advantages of hafnium compared to other rod materials is its strength and resistance to corrosion. Unfortunately in a fairly large reactor the cost of hafnium rods can be $1 million or more.

TANTALUM
Atomic number: 73
Chemical symbol: Ta
Group VB Transition Element

Tantalum is an extremely hard and very heavy metal. Its chemical inertness makes tantalum highly resistant to attack by substances in the human body. This has lead to a host of applications in dental and medical surgery. Tantalum as an alloying agent contributes corrosion resistance, ductility, hardness and a high melting point to a variety of other metals. Yet another major use of tantalum is in the construction of small but powerful electrolytic capacitors. These capacitors are specially useful in the miniaturized electronic circuitry that lies at the heart of such devices as cellular phones and computers.

TUNGSTEN
Atomic number: 74
Chemical symbol: W
Group VIB Transition Element

One of the most important uses of tungsten is in the manufacture of filaments for the common light bulb. Tungsten has the highest melting point -3410 degrees C and highest boiling point 5900 degrees C – of any metal. The high temperature applications of tungsten range from heating elements in electric heaters to the nozzles on the rocket motors of space vehicles. Electricity flowing through a coiled wire of tungsten produces enough heat to make the wire white hot. To prevent the metal from overheating inert gases such as nitrogen and argon are enclosed in the bulb containing a tungsten filament.

RHENIUM
Atomic number: 75
Chemical symbol. Re
Group VIIB Transition Element

Rhenium one of the rarest of elements was discovered in platinum ores by German chemists Ida Tacke, Walter Nodack and Otto Carl Berg in 1925. It is an extremely dense metal with a silvery gray luster and a melting point exceeded only by tungsten and carbon. This is the basis for rhenium's use in combination with tungsten to make thermocouples for measuring temperatures as high as 2000 degrees C. Rhenium is chiefly used as an alloying agent for fabricating metals that are resistant to wear such as those required for electric switch contacts and electrodes.

OSMIUM
Atomic number: 76
Chemical symbol. Os
Group VIIIB Transition Element

Because the pure metal is difficult to make, osmium is often fabricated as a powder which is then formed into solid mass by heating. The powder oxidizes in air and is slowly emitted as a strong smelling toxic gas capable of causing lung and skin damage. The emission of its poisonous oxide gas makes the use of osmium metal impractical. As an alloying additive however it is quite safe and is chiefly used to make hard alloys with such metals as platinum and iridium. These alloys are used for electric switch contacts, phonograph needles and fountain pen tips.

IRIDIUM
Atomic number:77
Chemical symbol. Ir
Group VIII B Transition Element

Iridium is a brittle yellowish white precious metal. It is generally found in ores containing platinum or nickel. Separating it from these ores is a laborious and costly task that is justified only by the simultaneous recovery of platinum and nickel. The chief application of iridium is as an additive to platinum creating alloys that increase the hardness of the latter metal. Iridium's resistance to corrosion makes it also useful in the fabrication of items that require absolute purity such as hypodermic needles and rocket engines.

PLATINUM
Atomic number: 78
Chemical symbol. Pt
Group VIII B Transition Element (Precious Metal)

Many uses of platinum take advantage of its chemical stability and inertness. It is used in petroleum refining, dentistry, the ceramics industry, the electrical and electronic industries, and is highly prized in the making of jewelry. Platinum is also useful to the automobile industry. It assists chemical reactions that clean up exhaust coming from the engines of cars, converting carbon monoxide and unburned fuel into water and carbon dioxide. Additionally a bar of iridium –platinum alloy serves as the world standard for the kilogram, the basic unit for mass in the metric system.

GOLD
Atomic number: 79
Chemical symbol: Au
Group IB Transition Element (Precious Metal)

Gold is traded in commodities exchanges and the fluctuations in its price are considered as an index of the health of the economy. It is the most ductile and malleable of all metals. Because it is also one of the most unreactive, it can sustain its brilliant luster. In nature gold is usually found as a pure metal, often as nuggets or flakes. Its purity is measured as carats. Pure gold is said to be 24-carat gold. Because it is very soft, however, most gold jewelry is made of 18 carat gold.

MERCURY
Atomic number: 80
Chemical symbol: Hg
Group II B Transition Element

Mercury is the only metal that is liquid at room temperature and remains a liquid over a very wide and convenient range of temperatures. Some common household products that contain mercury are thermometers, barometers, thermostats, silent wall switches and fluorescent bulbs. Industrial applications of mercury include diffusion pumps and mercury vapour lamps that generate the bluish white lights from street lighting. Another useful property of mercury is its ability to dissolve other metals to form alloys known as amalgams. Dentists often use silver-mercury amalgam to fill teeth.

THALLIUM
Atomic number: 81
Chemical symbol: Tl
Group III A Post-Transition Metal

A common source of thallium is zinc and lead refining. This malleable and heavy metal is quite active and slowly corrodes in air. Thallium and its compounds are extremely toxic and there is evidence that it can induce cancer. Even contact with skin can be dangerous although in extremely low concentrations thallium has been used in the treatment of ringworms. Thallium sulfate is an odourless and tasteless poison that was formerly used to kill rats and insects but it has now been banned in several countries.

LEAD
Atomic number: 82
Chemical symbol: Pb
Group IV A

Lead is a highly malleable metal that can be easily worked to make utensils of all kinds. Lead coins and sculpture have been found in Egyptian tombs dating back to 5000 BC. It is largely used to make electrodes of lead storage batteries. Lead is also an important component of solder used for making electrical connections on the circuit boards in computers and television sets. Glass screens of TV sets contain lead to shield the viewer from radiation. In fact every TV set contains nearly half a pound of lead.

BISMUTH
Atomic number: 83
Chemical symbol: Bi
Group VA Post transition Metal

Bismuth is a white brittle metal that has a slight yellowish tinge. The compound bismuth subnitrate has been used as an antacid in the treatment of ulcers. Bismuth oxide is a popular yellow pigment used in cosmetics. Like water bismuth is one of the few substances that expands when it changes from liquid to solid. This property is used to make alloys whose volume remains constant when they solidify. Metals alloyed with bismuth can be used for casts and molds that retain their exact dimensions even when filled with molten metals.

POLONIUM
Atomic number: 84
Chemical symbol: Po
Group VI A Metalloid

The discovery of polonium by Marie and Pierre Curie in 1898 defines one of the great moments in the history of science leading to the modern concept of the atomic nucleus and an understanding of its structure. Polonium has 27 known isotopes and all of them are radioactive. The one most readily available is polonium 210 ,a silvery metalloid that is quite volatile and 100,000 times more toxic than cyanide. In radiological laboratories the isotope mixed with powdered beryllium is often used to produce large amounts of neutrons without the use of nuclear reactor.

ASTATINE
Atomic number:85
Chemical symbol: At
Group VII A The Halogens

Small quantities of astatine exist naturally as the decay products of uranium and thorium. Astatine was first produced in 1940 by a team of radiochemists by bombarding bismuth with alpha particles. Only about 1 millionth of a gram of astatine has actually been produced artificially and it is therefore not surprising that little is known about its properties. Its chemistry should be fairly similar to that of iodine although there is some evidence that it may be slightly more metallic.

RADON
Atomic number: 86
Chemical symbol: Rn
Group VIII A The Noble Gases

Radon is produced as one of the by products of the radioactive decay of uranium and thorium. Radon-222 ,its longest lived isotope is found in substantial concentrations s a gas in soil because trace amounts of uranium are present in the Earth's crust. While it is growing, tobacco is subject to contamination by radon from the soil and the uranium rich phosphate fertilizers used by planters. When the tobacco in a cigarette is burned, the inhaled smoke subjects the smoker to levels of radiation 1000 times higher than those encountered by a worker in a nuclear power plant.

FRANCIUM
Atomic number: 87
Chemical symbol: Fr
Group I A The Alkali Metals

Francium is the heaviest of the alkali metals and one of the most unstable known. All of its isotopes are radioactive yet even its longest lived isotope francium-223 has a half life of only 21 minutes. Of its 30 known isotopes, only francium 223 exists in nature. All of the other isotopes of francium are produced artificially in accelerators and nuclear reactors and are too unstable to be studied in any depth. The element was discovered in 1939 by Marguerite Perey working at the Curie Institute in Paris. It is named for the country in which it was discovered.

RADIUM

Atomic number: 88
Chemical symbol: Ra
Group II A –The Alkaline Earth Metals

Radium was discovered by Marie and Pierre Curie in 1898. For the discovery of radium and polonium, Marie Curie was awarded the Nobel Prize in chemistry. It was her second; she had shared the first with her husband and Henri Becquerel in 1903 for discovery of radioactivity.
Pure radium metal has a brilliant white colour and is so luminescent that it glows in the dark giving off a faint blue colour. Radium is used in many medical facilities to generate the radioactive gas radon which is used for cancer therapy.

ACTINIUM
Atomic number: 89
Chemical symbol: Ac
Group III B Transition Element (The Actinides)

Actinium is a radioactive element produced naturally by the radioactive decay of the long lived elements radium and thorium. Very small amounts of it have been produced artificially and it has a very limited commercial application. Its chemical properties resemble those of lanthanum. Also like lanthanum, it is the first in a series of elements called the actinides which are analogous to lanthanides. Like the rare earths, these elements add electrons to an inner orbital shell and consequently have similar physical and chemical properties.

THORIUM
Atomic number: 90
Chemical symbol: Th
Group IIIB Transition Element (The Actinides)

Thorium is a radioactive silvery white metal that tarnishes very slowly when exposed to air. Monazite sand some of which is found in Florida beaches can contain upto 10% thorium. Despite its radioactivity ,thorium and its compounds have several commercial applications. It serves as an efficient emitter of electrons for electronic devices. The brilliant light that its oxide emits while burning also makes it useful in fabricating certain portable gas lamps. Thorium 232 ,an isotope with a half life of 14 billion years shows great promise of becoming a source of nuclear energy in the future.

PROTACTINIUM
Atomic number: 91
Chemical symbol: Pa
Group III B Transition Element (The Actinides)

It is one of the scarcest and most expensive of all the naturally existing elements. Only a few hundred grams are available for study. This meager amount was largely produced in England some 30 years ago where it was extracted from 60 tons of ore at a cost of half a million dollars. Not much is known about its physical and chemical properties. It is a silver white metal with a bright luster that it loses very slowly in air through oxidation. It is also known to be very toxic.

URANIUM
Atomic number: 92
Chemical symbol: U
Group III B Transition Element (The Actinides)

Uranium is the last and the heaviest of the naturally occurring elements. Discovered in 1841, it was the first radioactive element to be identified. In the late 1930's through experiments with uranium German scientists Lise Meitner and Otto Hahn observed a process that was later recognized to be a nuclear fission. The ability of the neutrons released during the fission of the uranium nucleus to themselves split other uranium nuclei was quickly utilized by the scientists to create a self-sustaining chain reaction .When controlled, this reaction produces the energy we obtain from nuclear reactors. When uncontrolled it can create an atomic explosion.

NEPTUNIUM
Atomic number: 93
Chemical symbol: Np
Group III B Transition Element (The Actinides)

Neptunium was the first artificially produced transuranium element. Working at the cyclotron at the University of California at Berkeley in 1940, US physicists Edwin McMillan and Philip Abelson produced neptunium by bombarding uranium with neutrons. It is now known that trace quantities of neptunium d actually exist in nature as the result of the actions of neutrons in the uranium element. Currently 18 isotopes of neptunium have been produced all of them radioactive. The most important and the first to be produced was neptunium 237 with a half life of 2.1 million years.

PLUTONIUM
Atomic number: 94
Chemical symbol: Pu
Group III B Transition Element (The Actinides)

Plutonium has 15 known isotopes all of them radioactive .Plutonium 239 is the most important because it readily fissions when bombarded by thermal neutrons. Like uranium 235, the nuclei of its atoms split into two intermediate sized nuclei (called fission fragments) releasing large amounts of energy and producing more neutrons to sustain a chain reaction. Mixed with powdered beryllium, it is an effective source of neutrons for scientific work. Plutonium can be produced in huge quantities in nuclear reactors. Its abundance has made it the number one choice for nuclear weapons.

AMERICIUM
Atomic number: 95
Chemical symbol: Am
Group III B Transition Element (The Actinides)

It was discovered in 1944 by a team of chemists under the leadership of Glenn Seaborg.His team produced americium-241 ,one of the 14 known isotopes all of which are radioactive. Americium 241 is made in large quantities in nuclear reactors. The intense gamma rays it emits makes it very useful as a portable source of X-rays. It is also used in smoke detectors.

CURIUM
Atomic number: 96
Chemical symbol: Cm
Group III B Transition Element (The Actinides)

Curium is a silvery white metal that is very reactive. The first of its 14 known isotopes to be discovered was curium 242. Curium 242 and curium 244 have been used as sources of energy in remote areas. The radiation these isotopes emit can be converted into heat and then into electricity by thermoelectric devices. Although it has a relatively short half life, the power output of curium 242 is impressive i.e. about two to three watts per gram. These compact units are useful for pacemakers ,remote navigational buoys and space missions.

BERKELIUM

Atomic number; 97
Chemical symbol: Bk
Group III B Transition Element (The Actinides)

It was discovered at UC Berkeley in 1949 by a team consisting of George Seaborg, Stanley Thompson and Albert Ghiorso and was named after the town. They synthesized it using a cyclotron to bombard a sample of americium 241 with alpha particles .Using berkelium 249, it was possible in 1962 to produce 3 billionth of a gram of berkelium chloride. No commercial or scientific applications have yet been developed.

CALIFORNIUM
Atomic number ; 98
Chemical symbol: Cf
Group III B Transition Element (The Actinides)

It was discovered by a team of chemists using a cyclotron to bombard curium 242 with alpha particles. The isotope californium 252 named for the State of California spontaneously emits neutrons. Neutron sources are occasionally hard to come by .Either a nuclear reactor is required or some highly radioactive emitter of alpha particles such as plutonium must be mixed with beryllium powder. The discovery of an extremely portable neutron source suggests many possible applications for californium 252.It can easily be taken into the fields for the analysis of oil bearing layers of earth or for mining of gold and silver.

EINSTEINIUM
Atomic number: 99
Chemical symbol: Es
Group III B Transition Element (The Actinides)

Albert Ghiorso and his co-workers discovered this element in 1952 while investigating the debris of hydrogen bomb explosion in the Pacific.16 isotopes are known ,the most stable being einsteinium 254 with a half life of 252 days. Most of these isotopes have been produced in the High Flux Isotope Reactor at Oak Ridge National Laboratory in Tennessee by irradiating plutonium 239 with intense beams of neutrons.

FERMIUM
Atomic number:100
*Chemical symbol:*Fm
Group III B Transition Element (The Actinides)

Like einsteinium, Fermium was identified in 1952 by Ghiorso and co-workers in the debris of hydrogen bomb explosion in the Pacific. Isotopes of fermium named after Enrico Fermi are usually synthesized by subjecting elements such as uranium and plutonium to intense neutron bombardment. In a neutron rich environment ,an element such as uranium can undergo successive neutron capture often absorbing as many as 16-17 neutrons to produce the heavy transuranium elements.

MENDELEVIUM
Atomic number:101
Chemical symbol: Md
Group III B Transition Element (The Actinides)

The ninth artificial transuranium element named for Dmitri Mendeleyev was discovered in 1955 by a group of scientists under Albert Ghiorso .Continuing their search for ever-heavier elements the team used the cyclotron at Berkeley to bombard einsteinium 253 with alpha particles (helium nuclei) and eventually fabricated mendelevium 256. The small amounts made its identification very difficult . It is often said that this element was synthesized one atom at a time. Only trace amounts of mendelevium isotopes have been made and little is known of their chemistry.

NOBELIUM
Atomic number: 102
Chemical symbol. No
Group III B Transition Element (The Actinides)

In creating nobelium 254, Ghiorso and his colleagues bombarded a sample of curium 246 with carbon 12 ions using the Heavy Ion Linear Accelerator . 11 isotopes have so far been synthesized and all are radioactive .Nobelium 259 is the longest lived with a half life of 57 minutes. Named for Alfred Nobel ,it has been produced in quantities large enough to permit the study of its chemical and physical properties.

LAWRENCIUM
Atomic number: 103
Chemical symbol. Lr
Group III B (The Actinides)

Continuing their astonishing string of discoveries, the Berkeley scientists synthesized and isolated lawrencium in 1961 by bombarding a mixture of 3 isotopes of californium with boron 10 and boron 11 ions using Heavy Ion Linear Accelerator. The target weighed only a few millionth of a gram yet the team managed to manufacture lawrencium 258 with a half –life of 4 seconds. It was named in honour of Ernest O.Lawrence, the inventor of the cyclotron.

RUTHERFORDIUM
Atomic number : 104
Chemical symbol. Rf
Group IV B A Transactinide

A history of competing claims confused the naming of element 104. The team from Berkeley as well as a group from Russia claimed credit for element 104. The American claim won the day. It is named after the New Zealander Ernest Rutherford!

DUBNIUM
Atomic number: 105
Chemical symbol: Db
Group VB A Transactinide.

Disputed claims of its discovery have plagued element 105. In 1970 Ghiorso and his team at Berkeley bombarded californium 249 with heavy nitrogen 15 ions and positively identified the element which they named after Otto Hahn and obtained endorsement from American Chemical Society. However in 1997 the IUPAC decided t change the name to Dubnium. Its chemical and physical properties are unknown.

SEABORGIUM
Atomic number: 106
Chemical symbol. Sg
Group VI B A Transactinide

Like the other two disputed elements, the claim of discovery of element 106 along with the right to name it was a subject of dispute. In 1974, a Russian team declared that they had produced unnilhexium. Because experiments failed to confirm their result, their claim was in doubt. About the same time ,scientists at Berkeley reported the discovery of unnilhexium 263 after bombarding californium 249 with oxygen 18. In 1993, scientists at the Lawrence Livermore and Berkeley Laboratories repeated the experiment and confirmed the result. It was named in honour of Glenn Seaborg.

BOHRIUM
Atomic number: 107
Chemical symbol: Bh
Group VII B A Transactinide

In 1981 ,the creation of unnilseptium was announced by physicists working in Darmstadt ,Germany at the GSI. The team proposed the name nielsbohrium after Neils Bohr . Their research claims were confirmed in 1992 by the IUPAC. In 1997,they changed the name to bohrium.

HASSIUM
Atomic number: 108
Chemical symbol: Hs
Group VIII B A Transactinide

In 1984 a team lead by Peter Ambruster and Gottfried Munzenberg announced the discovery of unniloctium ,element 108. This was the same team that had synthesized bohrium. The name they proposed was hassium after haasia the Latin name for the German state Hesse. In 1992 the IUPAC confirmed the findings and the name. The chemical and physical properties are unknown.

MEITNERIUM
Atomic number: 109
Chemical symbol: Mt
Group VIII B A Transactinide

In 1982, the Darmstadt team announced the discovery of element 109 by bombarding bismuth 209 with high energy iron 58 ions . Incredible as it may seem only 3 atoms were created and they decayed in a matter of 3.4 thousandth of a second. They proposed to name it after Lise Meitner who had fist described nuclear fission along with Otto Hahn.

UNUNNILIUM
Atomic number: 110
Chemical symbol, Uun
Group VIII B A Transactinide

After almost 10 years international scientists working at GSI in Germany successfully created four or five atoms of a new element 110. Using a large accelerator to drive nickel atoms to high speeds they bombarded a thin foil of lead with these fast moving atoms of nickel. The new element quickly breaks apart and decays into lighter atoms .It was detected by the 4 alpha particles it emits during its decay process.

UNUNUNIUM
Atomic number: 111
Chemical symbol: Uuu
Group IB A Transactinide

The chemical properties of element 111 are not known. As it lies in the same column as gold and silver it is presumably a metal. After accelerating nickel atoms to high speeds German researchers bombarded bismuth with these fast moving nickel atoms. The identification of this element is significant as it supports the theory that there exists an 'island of stability' for elements close to element 114. The element has a half life about 8 times that of ununnilium.

UNUNBIIUM

Atomic number: 112
Chemical symbol: Uub
Group II B A Transactinide

On February 9,1996 GSI in Germany announced the creation of element 112 all credit to the international team under Peter Ambruster. They had bombarded zinc atoms which had been accelerated to high speeds with fast moving bullets of lead. During the collision a zinc atom managed to fuse with the lead atom.

UNUNQUADIUM
Atomic number: 114
Chemical symbol: Uuq
Group IB A Transcatinide

In 1999 a team of scientists at the joint Institute for Nuclear Research in Russia announced the creation of a new ultra-heavy metal. The team utilized a cyclotron to bombard plutonium 244 with a beam of calcium 48 nuclei. After some 40 days of bombardment, a calicium nucleus with 20 protons fused with plutonium nucleus with 94 protons producing an element with 114 protons. Although unstable it survived a relatively long time.

The resolve to find nature's hidden answers has not abated. The quest remains for the ever continuing search for new superheavy elements. The driving force behind this effort is the search for knowledge that will initiate a rich new field of study of the nuclear and chemical properties of the elements.

There is also a more utilitarian motivation for the search of elements that make up the island of stability. Many scientists believe for example that these new elements will form unusual materials with exotic properties never before seen. The answers being sought in this effort are of fundamental importance to our understanding of the universe.

www.ingramcontent.com/pod-product-compliance
Lightning Source LLC
Chambersburg PA
CBHW070734180526
45167CB00004B/1754